AI Made Easy for Parents

Additional copies may be ordered from the publisher for educational,
business, promotional or premium use.
For information, contact ALIVE Book Publishing at:
alivebookpublishing.com

ISBN 13
978-1-63132-261-7 Paperback

Library of Congress Control Number: 2025915473

Library of Congress Cataloging-in-Publication Data
is available upon request.

First Edition

Published in the United States of America by ALIVE Book Publishing
an imprint of Advanced Publishing LLC
3200 A Danville Blvd., Suite 204, Alamo, California 94507
alivebookpublishing.com

PRINTED IN THE UNITED STATES OF AMERICA

10 9 8 7 6 5 4 3 2 1

AI Made Easy for Parents

Embrace AI today to thrive in tomorrow's world

James McConihe

ABOOKS

Alive Book Publishing

Contents

Chapter One

Introduction

Embracing AI in Family Life

My high school freshman son insisted he needed his computer to complete his English essay. When I reviewed his work, I could immediately tell he hadn't written it, and I was furious. "This is cheating!" I declared, forcing him to delete everything and start over "the right way." He was angry, but I stood my ground. What followed was painful: He stared at a blank Google Doc for over an hour, unable to craft even a basic thesis statement.

Has your son or daughter ever asked you about ChatGPT, or maybe you've caught them using an AI tool for homework? If you're feeling uncertain about how to respond, you're not alone.

Watching him struggle, I realized he wasn't trying to cheat. He simply needed help generating concepts and organizing his thoughts. That's when I shifted my approach. "Let me show you how to use AI for brainstorming, not copying," I said, introducing him to one of the leading AI models to explore possible thesis directions and outline structures. From that point, he took ownership of his work, writing confidently once he cleared the biggest hurdle: getting started.

We're living through one of the most significant technological shifts since the internet transformed our world, and this time, it's artificial intelligence leading the charge. AI has quietly woven itself into nearly every aspect of our daily lives. From the voice assistant that sets your morning alarm to the smart thermostat that adjusts your home's temperature, from grocery apps that predict your shopping list to the search engines that anticipate your questions—AI is already here, working behind the scenes in ways both obvious and invisible.

This transformation is happening at unprecedented speed. Companies are investing billions in AI infrastructure and integration. Every industry conference now centers around artificial intelligence. But for parents, the most pressing question isn't about corporate strategies or technological capabilities—it's about what this means for our children and how we can guide them through this new landscape.

This book was born from my own experience helping my son navigate homework with AI assistance. Like many parents, I found myself in uncharted territory, trying to balance the incredible potential of these tools with concerns about academic integrity and responsible use. I realized that most parents are facing the same challenge: Our children are already using AI—often without us even knowing it—while we're still figuring out what it is and how it works.

The reality in schools reflects this disconnect. Most middle and high schools have responded to AI by blocking websites ending in ".ai" and flagging any homework that appears to use copy-and-paste text. Teachers who once checked for plagiarism from web sources now struggle to identify AI-generated content. While a few progressive schools are developing thoughtful AI policies, the majority operate in gray areas, leaving students and parents to navigate this new terrain largely on their own.

Here's what should grab every parent's attention: Recent studies indicate that 65% of today's students will eventually work in jobs that don't yet exist, with AI literacy becoming as fundamental as reading and writing. This concept is discussed in the article titled "From-classroom to career: Building a future-ready global workforce" published on December 16, 2024 By the World Economic Forum on webforum.org.[1] Yet 53% of parents who are aware of AI haven't had a single conversation about it with their children. Read more about this in an article called "Generative AI in education" on internetmatters.org.[2]

Meanwhile, the workforce is already shifting. Employers increasingly expect both current employees and new hires to understand how to work with AI effectively. Those who don't adapt risk are being replaced by those who do. This isn't a distant future scenario—it's happening now.

Given this reality, we face a choice: We can fear AI and try to shield our children from it, or we can embrace it thoughtfully and teach them to use it responsibly and ethically. I believe the latter is not just the better option—it's the only realistic one. Parents who fear AI will raise children who are controlled by it; parents who embrace AI will raise children who control it.

---

1. https://www.weforum.org/stories/2024/12/from-classroom-to-career-building-a-future-ready-global-workforce/

2. https://www.internetmatters.org/hub/research/generative-ai-in-education-report/

In the chapters ahead, we'll start with AI basics explained in parent-friendly terms, then explore practical applications for homework and learning. The AI revolution is already underway, with or without our participation. The question isn't whether your family will interact with AI—it's how thoughtfully and intentionally you'll embrace it. Consider this book your roadmap to harnessing AI as a positive force in your household while equipping your children for the future they'll help create. By the end of this book, you will transform from an anxious AI bystander into your family's confident AI guide, equipped with the knowledge, tools, and mindset to raise children who understand that AI should work for them, not the other way around.

Chapter Two

Today's AI Landscape

In a nutshell, Artificial Intelligence (AI) allows computers and machines to mimic human thinking by copying processes like reasoning, learning, and solving problems. I first heard of AI growing up in the 1970s and 1980s. During that period, I was first introduced to the HAL 9000 in the 1969 film *2001: A Space Odyssey*, in which a spacecraft computer develops self-preservation instincts with deadly consequences. That was a creepy movie—or maybe it was just HAL's voice that creeped me out.

Later, the movie *Star Wars* (1977) came out with a comical and distinctive robot duo, R2-D2 and C-3PO. During the 1980s, AI robot movies and cartoons took off. In the 1980s, I loved watching the reruns of the early 1960s TV show *The Jetsons*, with their maid Rosey the Robot, who was sassy, would often get into trouble, and had to clean up after everyone. One of my favorite films was *WarGames* (1983), which featured WOPR/Joshua, a military supercomputer that nearly triggers nuclear war. One of the most popular movies was *The Terminator* (1984), which presented Skynet and its cyborg assassins as existential threats to humanity, solidifying the "AI rebellion" narrative. *RoboCop* (1987) explored the integration of human and machine, and *Star Trek: The Next Generation* (1987) featured Data, who is a self-aware and anatomically fully functional male android. As we moved into the Digital Age (1990s-2000s), more sophisticated and diverse AI films emerged, such as: *Terminator 2: Judg-*

ment Day (1991), *Resident Evil* (1996), *The Matrix* (1999), *Bicentennial Man* (1999), *A.I. Artificial Intelligence* (2001), *I Robot* (2004), *WALL-E* (2008), *Iron Man* (2008), and *Ex Machina* (2014). There are many more, but you get the point.

✦✦

AI is the broad term coined by John McCarthy in 1956 for describing the science of making machines intelligent. This dynamic and evolving discipline is defined by its ability to simulate human intelligence through machines. The area continues to advance, driven by technologies like machine learning and deep learning, while facing challenges in standardization and ethical considerations. These days, every major tech company offers its own AI chatbot or LLM (Large Language Model) – these tools require text input since they are language models, with new versions releasing quarterly or biannually. While there are constant benchmarks comparing different models, below are the top LLM AI chatbots most commonly used:

ChatGPT (OpenAI)

Pricing: Free limited use; $20/month (Plus), $200/month (Pro)
Website: chatgpt.com
Key Feature: Advanced generative AI with multimodal capabilities, custom GPTs, project-based contextual chats, real-time information access, team collaboration tools, versatile use cases, a user-friendly interface, and continuous model updates—making it a powerful and accessible AI assistant.

Claude (Anthropic)

Pricing: Free limited use; $20/month (Pro), $100–$200/month (Max)
Website: claude.ai
Key Feature: Excels in advanced reasoning, large context processing, vision analysis, coding, web search, ethical AI design, multilingual support, collaborative artifact creation, low hallucination rates, and scalable models optimized for safety and reliability.

Perplexity

Pricing: Free limited use; $20/month (Pro)
Website: perplexity.ai
Key Feature: Versatile research platform offering access to multiple AI models, deep research capabilities, advanced file analysis, personalized experiences, predictive analytics, and team collaboration tools for enhanced decision-making.

Grok (xAI)

Pricing: Free limited use; $8/month (X Premium), $30/month (SuperGrok)
Website: x.com
Key Feature: Real-time X platform integration with dual interaction modes, advanced reasoning, image and document processing, DeepSearch and Think features, multilingual voice capabilities, and edgy, unfiltered conversational responses.

Gemini (Google)

Pricing: Free limited use; $19.99/month (Pro), $249.99/month (Ultra)
Website: gemini.google.com
Key Feature: Multimodal AI processing text, images, audio, video, and code, offering advanced reasoning, coding support, and deep Google Workspace inte-

gration with features like Deep Research, Canvas, Audio Overview, and Gemini Live for real-time voice interactions.

Copilot (Microsoft)

Pricing: Free limited use; $20/month (Pro), $39/month (Max)
Website: copilot.microsoft.com
Key Feature: Embedded within Microsoft 365 apps, delivering real-time content generation, task automation, data analysis, and meeting summarization, with enterprise-grade security, compliance, and privacy protections.

Most LLM AI chatbots function similarly:

- They're accessible through a webpage

- They offer free access with limitations, typically requiring a subscription ($20/month) for advanced features

- You interact by typing questions or commands into a text field

- You can upload various documents (PDFs, Word files, etc.) for analysis

- They include research or search capabilities to find current information

- Many offer 'Think' or 'Reasoning' modes that show step-by-step thought processes

- Different models are available for different tasks

- Projects can be organized in folders

Example:

How can I help you today?

+   ⇅   ⊙ Research BETA   Claude Sonnet 4 ⌄

⌀ Write   ⬡ Learn   </> Code   ⎄ Life stuff   ⬚⬚ Connect apps

I recommend experimenting with these LLM AI chatbots to get a feel for how they function before diving into more substantial work. While free versions have limitations, consider subscribing to the Pro/Plus tiers on a monthly basis, rotating between different models since each offers unique capabilities. Think of it like TV streaming services—you switch between platforms based on which new series just launched. Don't overlook the research features—they're remarkably powerful, functioning like enhanced Google searches that deliver genuinely relevant results.

Image Creation (Text-To-Image):

The digital image creation landscape has exploded with options across platforms, making it easier than ever to create stunning visuals regardless of your skill level. Here are a few examples of image creation prompts:

> Create an image of a child blowing out candles on a birthday cake at home, surrounded by smiling parents and a sibling, balloons and festive decorations, bright storybook style

> Create a cartoon picture of attached photo

> Create an image of two young children playing with colorful building blocks on a sunny living-room floor, a smiling mother nearby, bright cartoon illustration style

You can upload a picture and ask it to convert it to a cartoon or add something to a photo. To improve your prompting, try uploading a photo and asking the AI chatbot to describe it to you. That will give you an idea of what it looks for when creating an image. Also, there are now guardrails that prevent generation of inappropriate or child-related explicit images.

DALL·E 3 (OpenAI)

Pricing: Paid (via ChatGPT Plus); limited free access via Bing
Website: openai.com/dall-e-3
Key Feature: Photorealistic text-to-image generation with advanced prompt handling and in-chat editing.

Grok (xAI)

Pricing: Paid (X Premium)
Website: grok.com
Key Features: Generates realistic portraits, art, and memes using Aurora, xAI's text-to-image model, integrated with the X platform.

Freepik

Pricing: Freemium (limited free images; paid tiers available)
Website: freepik.com
Key Features: AI-powered design platform offering templates, stock images, and creative tools for content creation.

Leonardo AI

Pricing: Freemium (limited free images; paid tiers available)
Website: leonardo.ai
Key Features: Real-time image generation with support for scribble-to-image, upscaling, and image-to-video creation.

Canva AI

Pricing: Freemium (free basic version; Pro subscription available)
Website: canva.com
Key Features: Beginner-friendly AI design tool with templates, drag-and-drop editing, and social media-ready graphics.

Adobe Firefly

Pricing: Freemium (free tier; full access with Adobe Creative Cloud subscription)

Website: firefly.adobe.com

Key Features: Professional text-to-image generation trained on Adobe Stock and integrated with Creative Cloud apps.

Midjourney

Pricing: Paid (starting around $10/month)

Website: midjourney.com

Key Features: Creates artistic, stylized images with multiple variations per prompt and supports remixing and upscaling.

WOMBO Dream

Pricing: Freemium (limited free images; paid tiers available)

Website: dream.ai

Key Features: Simple text-to-image generator designed for quick artwork creation with minimal learning curve.

Voice Generation (Text-To-Speech):

Voice generation is one of my favorite AI offerings, as I use audiobooks a lot during my exercise walks or on drives longer than thirty minutes. The robotic voices of early e-readers have been replaced by natural sounding voices, often including celebrities. The technology can now handle inflections, pauses, and different tones when reading – it is just amazing. Students with dyslexia could use AI to convert text-based lessons into audio formats.

I have included a few common text-to-speech (TTS) tools depending on what you need them for. I am going to be biased and say Speechify is my all-time favorite TTS, as you can scan text from a book, upload PDFs, connect with Kindle, access it from multiple devices, and the variety of different voices is great.

Speechify

Pricing: Free plan and paid plans from $29/month or $138/year
Website: speechify.com
Key Feature: Popular for audiobooks and accessibility, offering 200+ voices in 60+ languages, plus AI dubbing and voice cloning.

Natural Reader

Pricing: Free plan and paid plans from $20/month or $119/year
Website: naturalreaders.com
Key Feature: Audiobook and e-learning tool offering 200+ AI voices in 50+ languages, suitable for YouTube and commercial use.

ElevenLabs

Pricing: Free tier and paid plans starting at $5/month
Website: elevenlabs.io
Key Feature: Known for lifelike voices, supports text-to-speech and voice cloning in multiple languages for content creators and developers.

NotebookLM

Pricing: Free; pro version $19.99/month (Google AI Pro), free for students
Website: notebooklm.google
Key Feature: Creates audio overviews similar to podcasts from uploaded files or documents.

LOVO (Genny by Lovo AI)

Pricing: 14-day free trial; paid plans from $24/month or $288/year
Website: lovo.ai
Key Feature: Ideal for videos and podcasts, with voice cloning and team collaboration, offering 500+ voices in 100 languages.

Murf AI

Pricing: Free plan; paid plans start from $19/month or $228/year
Website: murf.ai
Key Feature: Popular for e-learning and corporate videos with 200+ voices in 20+ languages, plus API integration.

Text Generation (Speech-To-Text):

Speech-to-text technology brings real benefits to people's daily lives and work. For those who have trouble using keyboards or seeing screens clearly, speaking to create text makes computers much more accessible and opens up new possibilities for getting things done. Most of us can talk faster than we type, which means you can get your thoughts down quickly whether you're taking meeting notes, writing emails, or working on documents. It's especially handy when your hands are busy - like when you're driving and need to send a message, or cooking while dictating a grocery list. The technology really shines when it comes to capturing spoken content automatically. Instead of frantically scribbling notes during meetings or interviews, you can stay engaged in the conversation while the software handles the transcription. Many systems now work with multiple languages too, breaking down communication barriers in international settings.

I am seeing more and more people use speech-to-text technology to record conference keynotes, lectures, and panel discussions, and then process the text for summarized notes. Companies are finding it saves money on documentation tasks. Rather than hiring someone to transcribe customer calls, legal depositions, or medical records, the software can handle much of this work. Conference systems such as Teams, Zoom, and Google Meet now have a transcribe option in which they can summarize a meeting and list out action items. Live captioning at events or during video calls helps people who are deaf or hard of hearing participate fully in conversations. The transcribed text also becomes valuable data that businesses can analyze to understand customer patterns, improve services, or conduct research. What used to be time-consuming manual work now happens automatically, letting people focus on more important tasks.

Here are a few speech-to-text tools including a wearable option:

Microsoft Word Dictate

Pricing: Paid plans from $9.99/month or $99.99/year, varies by subscription
Website: microsoft.com/en-us/microsoft-365/word
Key Feature: Speech-to-text in Microsoft Word with real-time transcription, automatic punctuation, and voice commands for editing in multiple languages.

Google Docs Voice Typing

Pricing: Free for anyone with a Google account
Website: docs.google.com
Key Feature: Free built-in voice typing in Google Docs (Chrome only), supporting real-time transcription, multiple languages, and basic editing commands.

Otter AI

Pricing: Free plan and paid plans from $16.99/month or $99.96/year
Website: otter.ai
Key Feature: Real-time transcription with speaker identification, searchable transcripts, and integrations with Zoom, Google Meet, and Microsoft Teams.

Plaud

Pricing: ~$160 device + subscription from $9.90/month or $99.99/year and up
Website: plaud.ai
Key Feature: Wearable AI device for transcription and summarization in 60+ languages, featuring high-quality audio capture and cloud storage.

Video Creation (Text-To-Video):

AI video creation has transformed how we produce videos, making it faster and more accessible than ever before. This technology allows users to generate videos from simple text prompts or images, bypassing traditional production methods that require extensive equipment, expertise, and time. At its core, AI video generation combines machine learning with creative tools to produce content that would have once required an entire production team. AI video generators work by understanding and interpreting prompts. When you type something like "cat swimming in water," the AI uses its training to visualize and create that scene.

AI video creations are being used for various purposes:

- Creating educational content and explainer videos

- Developing marketing materials and advertisements

- Producing social media content

- Making short films and creative projects

- Converting text or image-based content into video format

AI video creation represents a significant shift in how we produce visual content, making video production more accessible to everyone regardless of their technical skills or resources. While the technology is still evolving, it already offers impressive capabilities that can save time and expand creative possibilities. This area of AI is rapidly growing so do expect great enhancements in the next couple of years.

Below are a few text-to-video tools:

Sora (OpenAI)

Pricing: $20/month (Plus via ChatGPT)
Website: sora.chatgpt.com
Key Feature: Generates high-resolution videos up to 20 seconds with asset integration, style presets, and editing tools like Remix and Storyboard.

Gemini Veo

Pricing: $19.99/month (Pro), $249.99/month (Ultra)
Website: gemini.google/overview/video-generation/
Key Feature: Produces 8-second, 720p videos from text/image prompts with native audio generation, cinematic realism, and character dialogue.

Runway/Aitubo

Pricing: Free tier and paid plans from $15/month
Website: app.aitubo.ai
Key Feature: AI research and media platform for video and image generation, popular among filmmakers, creators, and designers.

Freepik

Pricing: Freemium (limited free images; paid tiers)
Website: freepik.com
Key Feature: Creative design suite evolved from a stock image platform, offering templates and AI-generated images.

Vimmerse

Pricing: Paid plans from $2/120 credits to $120/10K credits
Website: vimmerse.net
Key Feature: Generates stunning images and transforms them into animated scenes or visual storytelling content.

Pollo AI

Pricing: Free plan; paid plans from $20/month or $119/year
Website: pollo.ai
Key Feature: All-in-one platform for generating high-quality AI videos and images from text prompts, static images, or video clips.

Music Generation (Text-To-Music):

AI music tools make it incredibly easy for parents and kids to create professional-sounding songs without any musical training. Simply type in what kind of music you want (like "upbeat pop song" or "calm study music") and these tools generate complete tracks in seconds. There are so many faucets in this music area that I will list a few top tools in Text-to-Music Software and AI MIDI Generators skipping areas in audio editing and remixes.

Below are a few text-to-music tools:

Suno AI

Pricing: Free plan; Pro ~$10/month (~500 songs)
Website: suno.com
Key Feature: Lyric-to-song generator with broad genre support, built-in voice and instrument layers, and high-quality outputs.

Udio

Pricing: Free plan; Pro ~$10/month (~500 songs)
Website: udio.com
Key Feature: Text-to-music and audio-to-audio generator that preserves the original audio style, ideal for co-creating new tracks.

Google MusicFX

Pricing: Free (experimental)
Website: labs.google/fx/tools/music-fx
Key Feature: Advanced text-to-music generator creating multi-minute compositions, trained on Google's massive music datasets.

Mubert

Pricing: Per track licensing (~$19–$499 for a 45-second track)
Website: mubert.com
Key Feature: AI text-to-audio generator producing royalty-free music using loop libraries and large datasets, ideal for podcasts and videos.

Soundraw

Pricing: $19.99/month or $203.88/year
Website: soundraw.io
Key Feature: Beat/song generator for content creators, offering royalty-free music based on mood, genre, and instruments.

SoundGen

Pricing: Free plan; paid plans $10–20/month or $99–199/year
Website: soundgen.io
Key Feature: Instrumental generator from text or images, supporting audio style transfer and remixing to vary arrangements.

Riffusion

Pricing: Free
Website: riffusion.com
Key Feature: Text-to-music generator producing instrumental and vocal tracks from prompts, with mobile support for photo-to-song creation.

Boomy

Pricing: Free plan; Premium $2.99/month; Pro $9.99/month
Website: boomy.com
Key Feature: Community-driven AI song creator for making and sharing original music, with options for royalties and social features.

Here are two AI MIDI Generators that are cutting-edge software harnessing AI to create MIDI (Musical Instrument Digital Interface) files, which digitally encode musical notes and parameters, empowering musicians, producers, and creators to effortlessly generate, tweak, and explore innovative music ideas, saving time and boosting creativity like never before.

HookPad Aria

Pricing: ~$14.99/month (annual discounts available); $199 lifetime
Website: hooktheory.com/hookpad/aria
Key Feature: AI songwriting assistant for generating context-aware chords and melodies in HookPad DAW, trained on large MIDI datasets.

Lemonaide

Pricing: $4.99–$9.99/month (VST plugin)
Website: lemonaide.ai
Key Feature: AI MIDI generator that creates melody and chord ideas, with drag-and-drop piano-roll editing and features for beatmakers.

Voice Assistants:

AI voice assistants have evolved significantly, with Apple's Siri at the forefront of integrating advanced technologies to enhance user experiences. These assistants have become integral to daily life—helping with tasks such as setting reminders, controlling smart home devices, and providing hands-free support while driving through systems like Apple CarPlay. Personally, I use a voice assistant in the car to read and respond to text messages, make phone calls, and navigate to different destinations.

More recently, voice assistants have been integrated with powerful AI models like OpenAI's ChatGPT and Anthropic's Claude, enabling more detailed responses and multimodal interactions.

The most popular voice assistants include:

- **Siri (Apple):** Launched in October 2011, known for seamless integration within the Apple ecosystem.

- **Alexa (Amazon):** Introduced in November 2014 with the Echo speaker, widely used for smart home control and third-party device compatibility.

- **Google Assistant (Google):** Released in October 2016 on the Google Pixel, recognized for its conversational capabilities and tight integration with Google services.

- **Bixby (Samsung):** Debuted in March 2017, primarily designed for Samsung devices.

Chapter Three

AI in Education

I believe it is critical for parents, and for students starting in elementary school to begin learning how to use AI now. Anthropic's CEO Dario Amodei recently stated that AI could wipe out half of all entry-level, white-collar jobs and spike unemployment to between 10-20% in the next one-to five-years. Read about this in the May 28, 2025 article "Behind the Curtain: A white-collar bloodbath" by Jim VandeHei and Mike Allen on axios.com.[1] Schools might want to police AI usage for fear of cheating, but the reality is different. Once these students enter the workforce, they will be expected to use and master AI to do their jobs efficiently as a basic requirement. Many college students now heavily use AI tools like ChatGPT to complete assignments, often without authorization. Read more about this in the May 27, 2025 article "Does College Still Have a Purpose in the Age of ChatGPT?" (Bloomberg Opinion) on livemint.com. [2] However, this book focuses on how to embrace AI rather than be discouraged by it. This isn't a trend to wait out; it's the new foundation of how work and learning operate. Those who master it early will have the advantage. We will discuss how to claim that advantage in future chapters.

---

1. https://www.axios.com/2025/05/28/ai-jobs-white-collar-unemployment -anthropic

2. https://www.livemint.com/education/news/does-college-still-have-a-pur pose-in-the-age-of-chatgpt-11748341533221.html

Since an AI chatbot or LLM requires text input, you can use raw text, PDFs, Word documents, or even images. While digital files are convenient, there are times when you will need a scanner to convert physical documents. Many households already have a printer with a built-in scanner, or you can use your mobile phone's scanning features. Personally, I prefer using a professional book scanner because it is much faster, and I do not have the patience for slow scanning apps or the warped text that sometimes results from using a phone or printer. Under fair use guidelines, it is generally allowed to scan small portions of copyrighted material for purposes such as research, teaching, or commentary, provided it does not harm the market for the original work.

Czur Book Scanner

If you are considering a book scanner, I recommend looking at Czur scanners, available at czur.com or through Amazon. These scanners require you to turn the pages by hand, and they include a foot pedal to speed up the process. They also come with Optical Character Recognition (OCR) software, which makes the scanned text easier for AI models to process. Some people may prefer to find a digital version instead of scanning. However, I prefer working with books I already own and will often scan the entire book for convenience. You'll also frequently need the exact same edition your child is using at school rather than a different version you might find online.

For learning and homework, I recommend using Anthropic's Claude as the primary choice due to its large context window (currently around two hundred thousand words, meaning it can process and remember huge amounts of text at once), ease of use, and strong reasoning capabilities. OpenAI's ChatGPT is also an excellent alternative. If you need to upload an entire book of up to three hundred pages, I recommend Claude specifically, and suggest creating a project rather than using a regular chat window. Claude currently supports files up to approximately thirty megabytes, though these limits may change over time.

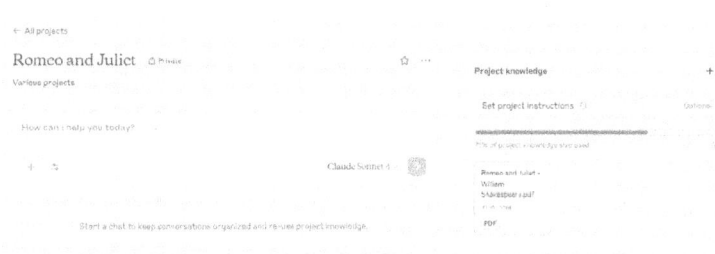

Claude Project View

Once your file(s) are uploaded into an LLM chatbot or project, you can do all sorts of things depending on what your needs are. Keep in mind—always double-check the output, as LLMs can sometimes be wrong.

Here are a few examples of what you can do with short articles, chapters, or assignments:

Create a study guide with key terms and detailed explanations and examples

Summarize chapter fifteen at a ninth grade level

What is chapter 10 about in detail?

Create a virtual flash card with this study guide

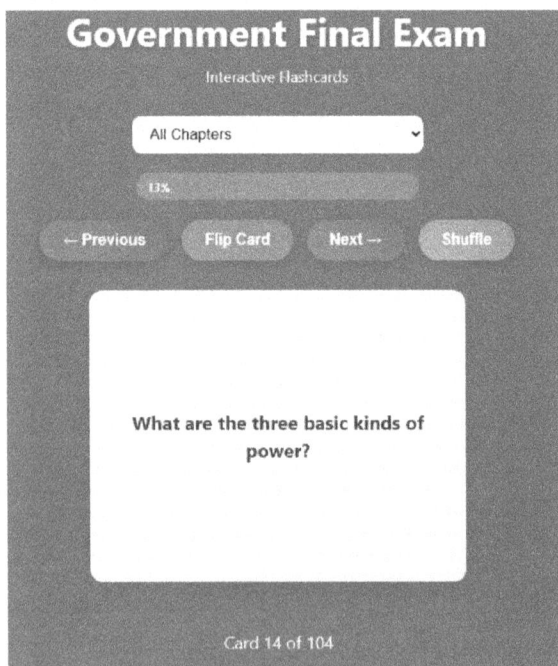

Government Final Exam

Interactive Flashcards

All Chapters

13%

← Previous   Flip Card   Next →   Shuffle

What are the three basic kinds of power?

Card 14 of 104

Claude can make virtual flashcards

I stay away from creating any tests, as from my experience, the questions are often never on the actual test, and your child might feel like they are wasting their time.

When working with an entire book, we will often use a few of these prompts:

List all and every character, sex if available, age if available, race, role, key elements of character and sorted by order of importance in entire book in a table format

What is this book about?

Summarize the setting, beginning, middle, climax and ending of book

Any contradictions?

What is the moral of the story?

Create a timeline of major events with dates

Create a family tree of the all the characters

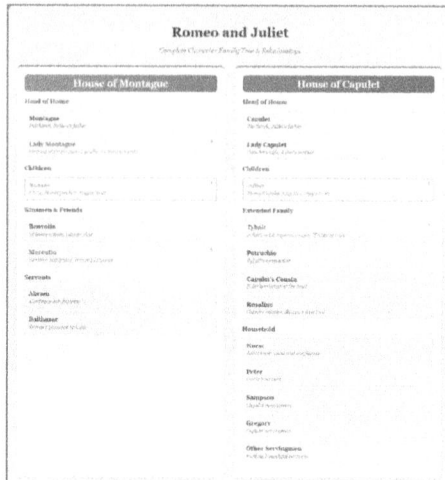

Family Tree Output Example

Depending on the book's complexity, I tell my child to pretend they can talk to the author and ask any question through the keyboard. This approach helped when our twelfth grader was reading *Into the Wild* by Jon Krakauer and wondered whether the main character's parents ever looked for him. The following prompt helped answer that question:

> Did Christopher McCandless's parents ever look for him or were worried about him?

If you need to write an essay about the book and are having a hard time coming up with a thesis statement, ask the LLM for possible ideas. Hopefully, the student will write their own essay. One simple but very powerful prompt we often use to check progress is:

> What do you think?

If you are paying for both Claude and ChatGPT and you ask the same prompt about a paper using both systems, the results are very interesting, as both LLMs will critique it with suggestions from different perspectives.

Don't forget to use Claude's Deep Research feature when appropriate. This tool functions like having a librarian, researcher, and writer all in one, working efficiently to answer complex questions or solve challenging problems. Deep Research gathers, combines, and explains information from many sources quickly and thoroughly.

Deep Research is fast, efficient, and comprehensive. It connects information from multiple sources, provides clear explanations, and can handle time-consuming research tasks. It's particularly great for:

- **Summarizing Big Ideas:** Reads numerous articles or studies and gives you the key points.

- **Understanding Rules or Laws:** Breaks down complicated laws or policies, like school regulations, into simple summaries.

- **Checking Out Businesses or Trends:** Pulls together information on companies or markets, like finding the best educational apps for kids.

- **Exploring History or Policies:** Explains causes and effects of events.

- **Diving into Technical Topics:** Simplifies complex subjects.

- **Verifying Facts:** Cross-checks sources to settle debates.

- **Writing Reports:** Creates detailed plans or guides.

Start experimenting with different prompts on their homework, as I am confident it will make learning more interesting and insightful. Reid Hoffman envisioned revolutionary possibilities when "every student on the planet had an AI tutor in their pocket, regardless of income." He saw AI as potentially closing educational gaps, noting that "a kid anywhere in the world will have access to a personalized digital tutor" without needing "a pile of cash to afford individualized instruction if their kid is struggling to pass algebra or chemistry." [3] In *AI Valley* by Gary Rivlin, Bill Gates similarly predicted that "in the next five-to-ten years, AI-driven software will finally deliver on the promise of revolutionizing the way people teach and learn." He believed AI could provide personalized tutoring that was previously only available to wealthy families.

---

3. *AI Valley* by Gary Rivlin

Chapter Four

Family Creativity with AI

Creating something from scratch doesn't have to start with staring at a blank page anymore. AI gives families different ways to jump into creative projects, depending on what clicks for each person. If your kids are visual thinkers, they can describe what they're imagining and watch it become a real picture. For families who love storytelling, you can team up with AI to build stories together—you suggest ideas, the AI adds to them, and before you know it, you've created something neither of you would have thought of alone. There are even apps that bring drawings to life or turn everyday objects around your house into parts of a game just by pointing your phone's camera at them. The nice thing is that whether your family is into pictures, art, music, stories, or writing, you can probably find an AI tool that fits what you already enjoy doing. Instead of everyone having to be good at the same thing, AI lets each person play to their strengths while still working on projects together—sometimes gathered around a screen, sometimes inspired to move away from it entirely.

As we explored in Chapter 3, the AI tools for images, videos, sound, and music have exploded over the last two or three years. These tools have been a game-changer—bringing both excitement and concern to those in the design business, while creating pure joy for everyday users who can now create virtually anything they imagine. Think about it: Parents can now whip up custom illustrations for their child's school project, design personalized birthday invitations, or even produce short videos for family memories without any design experience whatsoever. It's remarkable when you consider that tasks that once required expensive software and years of training can now be done in minutes with simple text prompts. Of course, this has significantly disrupted the creative world. Professional designers and artists are finding themselves in a landscape where AI can produce high-quality work in seconds, pushing them to adapt their skills and find new ways to stand out. Students, too, can now generate sophisticated visual content for assignments with ease. The speed and accessibility of these tools have fundamentally changed what's possible for anyone with a creative idea, regardless of their technical background.

✦✦

Creative Writing — LLMs can help you create amazing stories by coming up with cool characters, fun plot ideas, and even making you the hero of your own adventure. They turn writing into games where you can swap genres, rewrite endings, and build stories together with friends or family. These tools are also great homework helpers that can check your grammar, suggest better words, help organize essays, and even create silly stories using your spelling words to make studying way more fun. The best part is using AI like a creative buddy who gives you ideas to build on—you're still the real writer making the choices about what happens next, and you'll probably discover you're way more imaginative than you thought!

Here are some practical prompt examples you can try in each area:

Story Prompts and Imaginative Adventures

- **Personalized Story Prompts:** "Write a fun adventure story starring [child's name] and a talking [animal] who explore a hidden [magical place]."

- **Open-Ended Writing Prompts:** "Give me three imaginative writing prompts for a nine-year-old who loves dinosaurs and space."

- **Surprise Plot Starters:** "Start a story with a really surprising first line for a teen writer." / "I woke up to find my diary arguing with my homework."

Rewriting and Story Expansion Activities

- **Alternate Endings Challenge:** "Write an alternate ending where the villain becomes a friend."

- **Expand the Story:** "Here is my story draft. Can you expand it with more descriptive details and a surprise twist?"

- **Alternate Perspectives:** "Rewrite this scene from the viewpoint of the family dog." / "Retell the last paragraph as if it were a news report."

Idea Generation and Brainstorming with AI

- **Character and Setting Ideas:** "Give us five cool character ideas (name, age, personality) and five imaginative settings for a fantasy story."

- **Plot Twists and Story Ideas:** "What's an unexpected twist that could happen in my story about a lost puppy?" / "List three different ways a superhero story could end."

- **Titles and Outlines:** "Here's my story idea . . . .can you suggest ten chapter titles?" / "Help me outline a persuasive essay about recycling."

Style Transformation and Genre Play

- **Fairy Tale, Mystery, or Sci-Fi Makeovers:** "Rewrite this in the style of a fairy tale." / "Now rewrite it as a spooky mystery."

- **Mimic an Author or Style:** "Tell this story in the style of Dr. Seuss." / "Retell our story idea as if Shakespeare wrote it."

- **Tone and Vocabulary Swaps:** "Rewrite this paragraph to sound more formal and academic." / "Now make a version that's goofy and full of jokes." / "Simplify this story for a five-year-old reader." / "Add more advanced vocabulary to this description."

Interactive Writing Games and Challenges

- **Story Dice with AI:** "Give me three random things to include in a story (for example, an object, a character, and a place)."

- **Question-and-Answer Storytelling:** "What should our hero do next?" / "Choose an item from the room to help in the story." / "We're going to tell a story together. Ask my eight-year-old questions to help build the story, one step at a time." / "Our princess finds two doors in the castle. Does she open the red door or the blue door?"

- **Storytelling Challenges:** "Give us a first sentence for a story, and we'll see who can continue the story fastest in one minute." / "Every sentence must start with the next letter of the alphabet."

Educational Writing Support with AI

- **Essay and Homework Help:** "Help me outline a five-paragraph essay about climate change." / "What are some important points I should cover when writing about X?"

- **Grammar and Vocabulary Checker:** "Can you check this for grammar and suggest better word choices?"

- **Vocabulary-Building Stories:** "Create a mini story that includes all of the week's spelling words." / "Write a short adventure for a fifth grader that uses the words: courageous, glimpse, mysterious, tranquil."

- **Rewrite and Improve My Sentence:** "Make it better." / "Can you rewrite this to be more descriptive?"

Collaborative Storytelling with AI as a Co-Author

- **Parent-Child Co-Writing with AI:** "We have a hero and a villain; give us ideas for three exciting challenges they could face."

- **AI as a Creative Mediator:** "Our story has a dragon and a spaceship – how can we combine these into a cool ending?"

✦✦

Creating New images from Existing Ones

AI image generation works best when you start with high-quality, clear images and use detailed, vivid prompts. Describe the desired style, mood, lighting, and story elements you want the AI to create.

To get better results, experiment with cropping for improved composition and apply creative themes or narratives to transform ordinary photos. Try generating multiple variations and layering different effects to achieve more refined and imaginative results.

Here are a few examples of what you can do with tools such as OpenAI's DALL·E or other AI image generators when you upload a picture or graphic:

Create a realistic version of this drawing, paying attention to the details

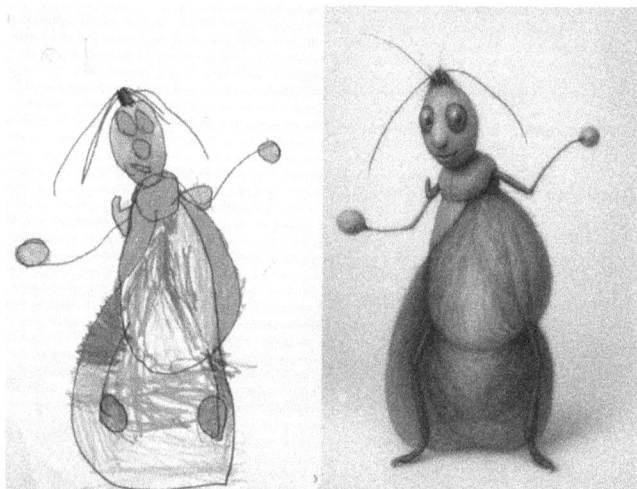

Create an image of what this would look like in Roman times

Remake this in Studio Ghibli style

Make a LEGO minifigure version of the person in the photo

Creating images from Text

Making these images can range from easy to very difficult depending on what you want to do. Remember, if you're stuck on text prompts, you can always upload an image and ask the AI to describe it, giving you ideas for better prompts. Creating effective AI art prompts means using clear, descriptive language that includes specific details about the subject, style, composition, and visual elements you want. Keep prompts concise and avoid contradictory terms that might confuse the AI. Research your chosen AI tool's capabilities and experiment with different keywords and artistic styles. Structure your prompts by first describing what the image is about, then adding necessary details, and finally specifying the desired artistic form and composition. Here are a few examples of what you can do with text-to-image tools such as OpenAI's DALL·E or other AI image generators from simple to detail descriptions:

Generate an impressionist-style oil painting of a stunning beach during sunset

16K realistic, intricate details, super closeup of a turquoise hummingbird filling the entire image, with detailed feathers, on a branch, dew drops.

A regal cat with ornate, baroque-style armor, standing proudly atop a mountain of glittering jewels in a dragon's lair. The cat's fur is a deep, luxurious black, contrasting with the gold and silver of the armor and the vibrant colors of the gems. The background features jagged rocks and a fiery glow emanating from the depths of the lair. Inspired by the works of Brom and Frank Frazetta, with a touch of Alphonse Mucha's decorative flair. Dark fantasy art, intricate details, dramatic lighting.

Creating Videos from Images and/or Text

AI video creation tools like Sora, Google's Veo, and RunwayML let you make short videos of up to twenty seconds per clip by simply describing what you want to see happen or uploading an image and adding motion prompts. Start with clear, simple descriptions of actions and camera movements. Then use the built-in editing tools to refine your results through iteration until you get the video you envision. This area of AI keeps improving rapidly, with major advances every six months or faster. Soon, creating short films or even full-length movies may become accessible to everyday users.

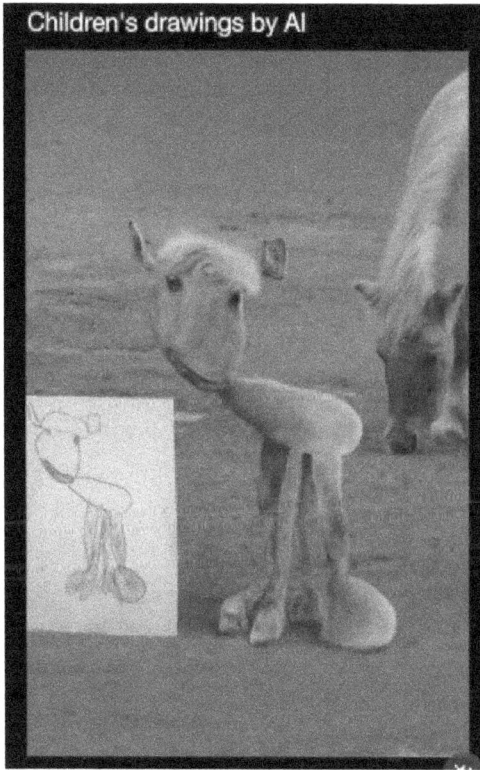

AI brings children's drawings to life

0:00 / 0:10

Flying through coral-lined streets of an underwater suburban neighborhood.

✦✦

Creating Music from AI

AI music tools have revolutionized music creation by allowing parents and kids to generate professional-quality songs without any musical knowledge or expensive equipment. Simply describe what you want in plain English (for example, "happy birthday song with guitar" or "relaxing piano music for bedtime"), and these platforms create complete tracks within seconds. These tools go beyond basic text-to-music creation. You can also convert images to music, transfer audio styles, work with MIDI inputs, train voice models, generate reference tracks, and even synchronize music to videos.

Popular user-friendly options include:

- **Suno AI** for turning text into full songs with vocals

- **Soundraw** for creating custom background music by selecting mood and genre

- **Boomy** for one-click song creation that can even be uploaded to Spotify and other streaming platforms

These tools also offer powerful features like separating vocals from instruments (perfect for karaoke), cleaning up home recordings, and even cloning voices to create personalized songs. Most platforms offer free versions with generous limits, making it easy for families to experiment with creating custom music for birthdays, school projects, family videos, or just for fun.

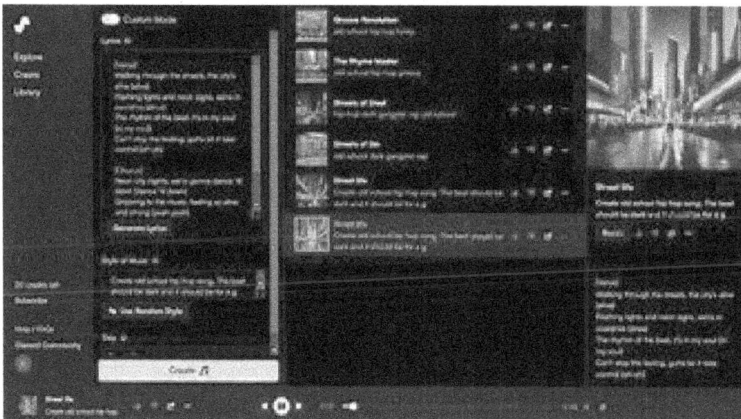

Screenshot of Suno AI

AI Storytelling Apps for Children

AI-powered storytelling platforms are revolutionizing creative expression for children by providing intuitive tools that help young minds translate their vivid imaginations into written stories and illustrations. Parent and reviewer Amanda Caswell recently tested five prominent AI storytelling applications and found each offer's unique features in a November 19, 2024 article titled "As a busy mom these are the 5 best AI Storytelling apps for kids I've tried" on tomsguid e.com.

- **Magic Story AI** — Allows children to become heroes in personalized stories using uploaded photos

- **StorySpark** — Guides kids through the creative process with prompts and mentorship

- **StoryBee** — A subscription-based platform that creates personalized stories but lacks visual engagement

- **BedtimeStory.ai** — Produces hyper-personalized bedtime stories with Disney-quality illustrations

- **OnceUponABot** — Emerged as the top choice for its ease of use, immediate story generation with illustrations, and free tier allowing three books per month

Chapter Five

Save Time and Money with AI

Artificial intelligence isn't just for tech experts anymore—it's become an incredibly powerful tool that busy parents can use to simplify daily life, save substantial money, and reclaim precious time. From finding discount codes that can save hundreds of dollars on family purchases to creating personalized learning experiences for your children, AI tools are quietly revolutionizing how families manage everything from grocery budgets to homework help. Whether you're planning a family vacation, tracking household expenses, or helping your teenager with algebra, the same AI technologies that power major corporations are now available as free or low-cost apps that require no technical expertise at all. Here are seven areas where AI can help you save money:

Shopping & Deals

- **Discount and Promo Codes**—Large Language Models (LLMs) like Perplexity, Grok, and ChatGPT (with research mode on) are great for finding discount or promo codes for online shopping, conferences, or events. By using these tools regularly, I've saved hundreds of dollars over the past year. Earlier this year, I even attended CES (Consumer Electronics Show) in Las Vegas for free. So, try really hard not to leave that promo code box empty when shopping!

- **Price Comparison**—With research mode enabled, AI excels at price comparisons, often providing ratings, as well as pros and cons to help you make informed decisions.

Education & Tutoring

- **Start Small and Save**—Begin by testing free tools first (ChatGPT, Claude, other LLMs, Khan Academy, and apps like Photomath) to find what works best for your child's learning style. Upgrade to paid versions only when you've fully utilized the free features and see clear additional value.

- **Smart Questioning Strategies**—Teach children to ask specific, targeted questions rather than vague ones: "How do I solve $2x + 5 = 17$?" vs. "I don't get algebra." Use AI to check work and explain mistakes rather than just getting answers. Encourage follow-up questions like "Can you give me another example?" or "Why does this work?"

- **Reading & Writing**—Get grammar and style feedback, replacing writing tutors. Use LLMs for reading comprehension help and essay brainstorming. Create custom vocabulary quizzes and flashcards with AI instead of buying materials.

- **Language Learning**—Use LLMs as free conversation partners: "Let's chat in Spanish and correct my mistakes." AI can role-play scenarios, eliminating the need for expensive conversation classes. The major LLMs are fluent in about ten-to-twenty languages.

- **Test Prep & Study Skills**—Create study guides and key terms or even practice quizzes. Use research mode to help with AP exam prep or college-level test prep.

Time Management

- **Free automated scheduling**—Reclaim.ai eliminates hours of manual calendar juggling by automatically finding and protecting time for both work and family priorities, with a completely free plan for individual users that handles unlimited tasks.

- **Free time tracking forever**—Clockify helps identify where your time actually goes (work vs. family tasks) to eliminate inefficiencies and time waste, with unlimited free tracking that costs nothing to use.

- **Free smart task management**—Todoist reduces daily planning overhead with AI that suggests optimal due dates and task priorities, plus free family task sharing that eliminates the mental load of remembering everyone's responsibilities.

- **Free habit-based scheduling**—Any.do learns your routines (e.g., grocery shopping, appointments) and suggests optimal scheduling, reducing decision fatigue and planning time with its free core features.

Chapter Six

AI-Powered College Prep

College preparation has never been more complex or more exciting. Between standardized tests, essay writing, financial aid forms, and choosing the right school, families today face an overwhelming maze of decisions that can determine their child's future. But here's the game-changer: artificial intelligence is revolutionizing every step of the college prep journey, transforming what used to be stressful, time-consuming processes into personalized, efficient experiences. From AI tutors that provide 24/7 homework help to smart scholarship finders that uncover hidden funding opportunities, from writing coaches that polish college essays to virtual college counselors that match students with their perfect schools, AI tools are leveling the playing field. These technologies give every family access to resources that were once available only to those who could afford expensive private counselors. This chapter will show you exactly how to harness these powerful AI tools to help your child not just survive the college prep process, but thrive in it, all while maintaining their mental health and discovering exciting possibilities they never knew existed.

Academic Preparation

- **Standardized Test Prep (SAT, ACT, AP)**—Standardized tests like the SAT, ACT, and AP exams often cause stress for students. AI-powered test prep tools are transforming how students practice and improve their scores. Many online prep programs now use AI algorithms to create adaptive tests, such as Khan Academy Official SAT Practice (khanacademy.org/digital-sat), and R.test (rtest.ai) which offers an AI-powered SAT/ACT diagnostic that predicts your score based on one full-length test and identifies areas to study. These tools exemplify how AI makes test prep more targeted and convenient, reducing the time students spend on material they already know while focusing practice on their problem areas.

- **Academic Support & Tutoring**—LLM-powered chatbots and AI tutors such as Khan Academy's Khanmigo (khanmigo.ai) are available anytime to answer academic questions across all subjects, from complex math problems to challenging science concepts. These AI tutors can break down difficult topics, provide step-by-step explanations, and offer practice problems tailored to your learning needs. Don't forget the virtual flashcards discussed in chapter four, which are very powerful for memorization and review. Here are some effective prompts to get the most out of AI tutoring:

How do I solve this calculus problem?

Can you explain this chemistry concept?

Application Process

- **Essay Writing & Refinement**—The college admissions essay is a critical piece of the application, and AI can be a helpful writing coach to help students brainstorm ideas, create outlines, polish grammar, and recast sentences. AI should help develop and refine a student's own story, not replace your unique voice and perspective, which ad-

- **Making Sure Computers Can Read Your Resume**—Applicant Tracking Systems (ATS) are computer programs that 75-99% of companies use to automatically screen resumes before any human ever sees them, which means your perfectly qualified child might never get a chance if their resume isn't formatted correctly. These systems scan for specific keywords from the job posting, check formatting, and rank resumes based on how well they match what the company is looking for, often rejecting good candidates simply because they used different words to describe the same skills. If your resume isn't ATS-friendly, it gets filtered out immediately and goes into a digital trash bin, no matter how impressive the experience or qualifications are. Tools like Jobscan help you check your resume against job postings to see your "match score" and suggest which keywords to add naturally throughout your resume. Think of ATS optimization as the entry ticket to getting your resume in front of a real person who can actually hire you. The key is weaving job posting terms naturally into your bullet points and using clean formatting with standard fonts like Arial, simple section headings, and avoiding graphics or fancy tables that confuse both computer systems and human readers.

Applications That Stand Out

- **Cover Letters Made Easy**—AI tools like ChatGPT and Claude can write personalized cover letters in minutes rather than hours by using a simple three-part formula: research the company and role, explain why your child is a good fit, and express genuine excitement about the opportunity. Instead of starting from a blank page, you can use prompts that give AI the specific information it needs, such as the job posting, your child's relevant experience, and what interests them about the company. The key to success is always editing the AI-generated letter to sound like your child's authentic voice rather than generic corporate missions officers value. AI excels at identifying awkward phrasing, suggesting sentence variety, and helping with tone adjustments to make writing clearer and more engaging. Whether you need help with formality, clarity, or overall flow, AI can provide targeted suggestions while preserving your authentic voice. Try these prompts to improve your college essays:

> Here's my draft paragraph—any tips to make it clearer and more engaging?

> Make this sentence sound more formal.

> Simplify this paragraph but keep the core idea.

- **Getting Feedback**—The reasoning capabilities of AI tools like ChatGPT and Claude have improved so much over the past year that they can now provide impressive feedback on documents and text. AI can provide general feedback on entire documents or help you understand what impression your essay creates, showing whether your key qualities are coming through effectively. While AI feedback might not be as detailed as a human counselor's, it can still spot clear problems, like a missing conclusion or a lack of personal touch. This helps students make revisions early on. Try these prompts to get useful feedback on your college essays:

> What do you think of this essay?

> What impression do you get of me from reading this essay?

- **College Search & Selection**—AI LLMs simplify college searches by acting like smart counselors, matching students to schools based on their grades, interests, and preferences. They analyze vast amounts of data, including admissions stats, programs, and campus life, to create personalized college lists, often suggesting schools families might not have considered. These tools compare colleges side by side, looking at costs, graduation rates, and majors, and predict admission chances to help build a balanced list of reach, match, and safety schools. Students can cut through information overload and find colleges, including hidden gems, that truly fit their goals. With research mode on, try a couple of these prompts:

> Help me create a balanced college list. My stats: [insert GPA/test scores], intended major: computer science, preferences: urban setting, diverse student body, strong internship/co-op programs, budget under $40k/year. I need 2-3 reach, 3-4 match, and 2-3 safety schools.

> Compare [College A] vs [College B] vs [College C] for a pre-med student in terms of: acceptance rates to medical school, research opportunities, class sizes, campus culture, and total cost of attendance. Present this in a clear format.

Financial Planning

- **Scholarship Discovery**—AI makes finding scholarships much easier by matching students with opportunities that fit their grades, background, major, and hobbies. Instead of spending hours searching through endless lists, AI tools create personalized lists of scholarships you actually qualify for, including hidden opportunities you might otherwise miss. Apps like ScholarshipOwl even fill out application forms for you and suggest similar scholarships to apply for. With AI, you can apply to more scholarships faster, increasing your chances of securing funding for school while saving significant time. With research mode on, try a couple of these prompts:

> I'm a high school senior with a 3.6 GPA, 1280 SAT, planning to major in nursing, involved in hospital volunteering and school orchestra, first-generation college student from California. Find scholarships that match my profile, including local, state, and national opportunities.

> Find scholarships for students with learning disabilities who excel in STEM fields. I have ADHD, 3.4 GPA, strong in math/science, and want to study biomedical engineering.

- **Financial Aid Navigation**—AI language models are like friendly financial aid coaches, instantly breaking down tricky FAFSA terms and rules into plain English. They can explain things like the difference between subsidized and unsubsidized loans, what Expected Family Contribution (EFC) really means, and how various types of aid work. Students can ask them to compare financial aid offers from different colleges, with the AI spelling out the real costs and helping families figure out which package is the best deal, considering grants, loans, and work-study options. Additionally, these models can walk families through filling out the FAFSA, answering questions about topics such as reporting income, dependency status, and what documents are needed, right when they're applying. There's no waiting for official help. It's like having a guide that makes financial aid easier to understand and navigate, leveling the playing field for everyone.

Try a few of these prompts for financial aid navigation:

> I'm a high school senior starting the FAFSA for the first time. Walk me through the basic steps and what documents I need to gather before beginning the application.

> My family owns a small business. How should we report business income and assets on the FAFSA? What expenses can we deduct?

> We have unusual circumstances: my parent is unemployed, we have high medical expenses, and I have a sibling in college. How do I explain this on the FAFSA or to financial aid offices?

College Exploration

- **Career & Major Exploration**—AI revolutionizes career and major exploration through interactive assessments that use machine learning to analyze students' hobbies, favorite subjects, values, and work preferences, then recommend tailored career paths. These tools dynamically adjust questions based on responses and suggest unique major combinations, such as bioinformatics for students interested in biology and computing. By incorporating labor market data, AI offers insights into job outlooks, salaries, and required skills, aligning paths with emerging fields. Advanced platforms provide career simulations and AI mentoring, allowing students to experience realistic job scenarios and interact with virtual professionals. Tools like CareerVillage's AI Career Coach, YesChat's Major Picker, ChatGPT, and Claude help students discover both obvious and niche careers, guiding informed major choices.

- **Campus Visits (Virtual & In-Person)**—AI transforms campus visits by creating personalized tour schedules that align with students' interests, such as coordinating engineering lab tours and soccer coach meetings in one day while optimizing travel between colleges. On campus, AI-powered mobile apps deliver tailored commentary about relevant buildings, match students with ambassadors from similar backgrounds, and provide instant chatbot answers during self-guided tours. For remote families, AI enhances virtual tours with interactive 360° views, chatbots like Arizona State's "Sunny" that offer campus life insights, and real-time representative chats. After visits, AI sends customized follow-up emails addressing specific questions and linking to relevant academic or research opportunities. This technology makes campus visits, whether virtual or in-person, more engaging and informative, helping students make better college choices.

Student Wellness

- **Mental Health Support**—AI supports students during college prep stress with 24/7 mental health chatbots like Wysa and Woebot. These tools use cognitive-behavioral therapy techniques to offer coping exercises and mindfulness activities, serving as a non-judgmental first step for teens hesitant to seek counselors. When functioning as an academic coach, AI identifies procrastination or time management issues, then suggests strategies like the Pomodoro Technique or sends motivational messages based on behavior patterns. It also connects students to peer tutors or human resources by detecting distress in school forums, fostering community support. AI-powered wellness apps like Calm provide tailored meditation and biofeedback tools to manage sleep and stress, though they cannot replace professional counseling. These tools complement human connection, reducing isolation and teaching healthy habits, but they lack the true empathy of human support and are not substitutes for professional care for serious mental health issues.

- **Balance and Mindfulness**—AI supports student balance and mindfulness through personalized meditation apps like Calm and Headspace, which recommend tailored meditations based on user feedback and stress patterns. AI-powered biofeedback tools monitor sleep and screen time, suggesting breaks when needed. AI wellness platforms detect stress via wearables or self-reports, then generate relaxing playlists or advise device disconnection during late-night study sessions. Time management tools like Reclaim.ai, Focus Booster App, and ClickUp use AI to create sustainable routines, prioritize tasks, and ensure rest, acting as digital wellness coaches. These AI tools adapt to individual needs, providing immediate, personalized stress-relief support when traditional resources are unavailable.

Chapter Seven

Getting Hired with AI

The job market has fundamentally shifted in ways that catch many parents off guard. Most companies now use AI to automatically screen resumes before any human ever sees them, which means your child's perfectly qualified application might never make it past a computer program. But here's the encouraging news: Job seekers who embrace AI tools are seeing significantly more responses from employers, proving that understanding these technologies isn't just helpful, but becoming essential for career success. Whether your child is a high school student applying for their first job, a college graduate entering a competitive market, or someone looking to change careers, AI tools can level the playing field by helping them create standout resumes, write compelling cover letters, optimize their LinkedIn profiles, and even practice interviews. This chapter will show you exactly how to guide your child through this new landscape, starting with free tools and simple strategies that can dramatically improve their chances of landing the job they want.

Getting Started with AI Job Tools

- **AI Job Search Basics for Parents**—AI tools like ChatGPT and Claude can help write resumes, create cover letters, optimize LinkedIn profiles, and even practice interviews by generating personalized content based on your child's experience and target jobs. Start with free versions of tools like ChatGPT, Teal, and Jobscan to learn the basics before considering paid upgrades that offer faster results and advanced features. Setting up accounts is straightforward since most require just an email and basic profile information, and you can organize everything using a simple spreadsheet to track applications and tool usage. Expect to spend one-two weeks learning the tools and seeing initial improvements in application quality, with most families noticing increased interview requests within four-to-siz weeks of consistent AI-assisted applications. The key is starting simple with free tools and gradually building skills rather than trying to master everything at once.

- **The Essential AI Toolkit**—Your core toolkit includes ChatGPT or Claude for writing resumes and cover letters, Jobscan or SkillSyncer to make sure your resume passes through company computer systems that filter applications, LinkedIn and tools like Careerflow for finding job matches, and Interview Warmup or Huru for practicing interviews safely. These tools work together like a complete job search team where each one handles a different part of the process, from creating great application materials to preparing for interviews. Most of these offer free versions that provide plenty of features to get started, so you can test what works best for your situation before spending money. The beauty of this toolkit is that once you learn the basics, you can use these same tools for every job application, making the whole process faster and more effective. Great Prompts to Try:

Write 3 resume bullet points for a [job title] position that show leadership and include specific numbers or results.

Analyze this job description and my resume. What keywords am I missing and how can I improve my fit?

Pretend you're interviewing me for [job title]. Ask me 5 common questions and give feedback on my answers

Creating Winning Resumes with AI

- **Writing Resumes That Get Noticed**—Start by collecting all your child's work experience, volunteer activities, school projects, and achievements in one place, then use AI tools like ChatGPT or Claude to transform basic job descriptions into compelling accomplishments that employers actually want to read. Instead of writing "Responsible for customer service duties," AI can help you reframe it as "Resolved 95% of customer inquiries within twenty-four hours, improving customer satisfaction scores by 28% and reducing escalation requests by 40%." The magic happens when you add specific numbers, percentages, and measurable results to every bullet point, which AI can help you brainstorm if you're not sure what metrics to include. Simple prompts can instantly upgrade boring resume content into impressive accomplishments by focusing on leadership examples and measurable outcomes. Remember to always review and personalize what AI suggests to make sure it accurately reflects your child's actual experience and sounds authentic to their voice. Try these prompts:

Write 3 resume bullets for a [job title] that demonstrate leadership and include specific results.

Transform this job duty into an achievement with numbers: [paste description].

speak, which means adding personal touches, adjusting the tone, and ensuring all details are accurate. Remember that cover letters should complement, not repeat, the resume by telling a story about why this particular job at this specific company makes perfect sense for your child's career goals. With practice, you can generate a solid first draft in under ten minutes and spend just a few more minutes personalizing it for each application. Here are a few prompts to test out:

> Write a cover letter for [position] at [company]. Here's the job description: [paste job posting]. Here are my relevant qualifications: [list three-to-four key experiences]. I'm excited about this role because [specific reason about company/role].

> Write a cover letter explaining how my background in [previous field] makes me a strong candidate for [new position] at [company]. Focus on transferable skills like [skill 1, skill 2, skill 3].

> Write a cover letter for an entry-level [position] that highlights my potential, eagerness to learn, and relevant experience from [internships/projects/volunteer work]. Make it enthusiastic but professional.

- **LinkedIn Profile That Gets You Found**—LinkedIn has become essential for job searching because recruiters actively use it to find candidates, making a well-optimized profile your job search tool that works even when you're not actively looking. AI tools like ChatGPT and Claude can help craft compelling headlines that include relevant keywords and write engaging "About" sections that tell your child's professional story in a way that both recruiters and LinkedIn's search algorithm will notice. The key is optimizing for search by including industry-specific terms and skills that recruiters typically search for, while making sure the content sounds natural and reflects your child's personality and career goals. A complete, keyword-rich profile with a professional photo and regular activity signals to LinkedIn's algorithm that this person is serious about their career, which boosts visibility in recruiter searches. Think of LinkedIn optimization as creating a professional billboard that showcases your child's best qualities and makes them easy to find when opportunities arise.

Try these prompts to better your LinkedIn profile:

Create five LinkedIn headlines for a [job title/student] with [X years] experience in [field] who wants to work in [target role]. Include these keywords: [list three-to-five relevant skills]. Keep each headline under 120 characters.

Write a compelling LinkedIn About section for someone with my background: [brief summary of experience/education]. Make it 150 words or less, include keywords for [target industry], and end with a call-to-action for networking. Use a conversational but professional tone.

Rewrite this job description for LinkedIn to focus on achievements rather than duties: [paste current description]. Include metrics where possible and use keywords relevant to [target field].

Write a brief LinkedIn connection request message to a [recruiter/professional] at [company name]. Mention that I'm interested in [specific role/field] and have experience in [relevant area]. Keep it under two hundred characters and professional but friendly.

Finding and Landing the Right Job

- **Smart Job Searching with AI**—AI-powered job matching platforms like LinkedIn, Careerflow, and Sonara can automatically find relevant positions based on your child's skills and preferences, saving hours of manual searching through job boards and surfacing opportunities you might have missed. These tools also help you stay organized by tracking applications, deadlines, and follow-ups in one dashboard, which is crucial when applying to multiple positions and prevents the embarrassment of duplicate applications or missed deadlines. The key is knowing when to use automated features versus adding a personal touch—use automation for initial job discovery and application tracking, but always customize resumes and cover letters for positions your child really wants. AI chatbots such as Grok, can also quickly research companies by summarizing their recent news, culture, financial performance, and growth trajectory, giving you talking points for interviews and helping you determine if it's truly a good fit.

- **Interview Preparation and Success**—AI tools like Interview Warmup and ChatGPT can simulate realistic interview scenarios, ask you common questions for your target role, and provide feedback on your answers to help build confidence and improve your responses before the real thing. These tools excel at quickly researching companies by summarizing their recent news, culture, values, and key business information, giving you relevant talking points and demonstrating genuine interest during interviews. Practicing with AI helps you prepare strong STAR method responses (Situation, Task, Action, Result) for behavioral questions and refine your explanations of technical skills or experiences until they sound natural and compelling. AI can also help you prepare thoughtful questions to ask the interviewer, draft professional thank-you emails after interviews, and even practice salary negotiation conversations in a low-pressure environment. The goal is using AI to practice until you feel confident and prepared, but then bringing your authentic personality and genuine enthusiasm to the actual interview.

Pretend you're interviewing me for a [job title] position at [company type]. Ask me five common behavioral questions one at a time, wait for my response, then give specific feedback on how to improve my answer.

Research [company name] and provide a summary of their recent news, company culture, main products/services, and key facts I should know for an interview. Also suggest three thoughtful questions I could ask the interviewer.

Help me create a strong STAR method response for the question 'Tell me about a time you overcame a challenge.' Here's my situation: [describe experience]. Make it compelling and include specific results.

Provide a detailed bulleted SWOT analysis of [company]'s positioning in their core business.

Advanced Strategies

- **Salary Negotiation and Career Planning**—AI has transformed salary negotiation from an intimidating guessing game into a confident, data-driven conversation. Tools like Payscale and Glassdoor now offer real-time compensation data and company-specific insights, while platforms like Salary Negotiator let you practice realistic negotiation scenarios with personalized feedback. Think of it as a flight simulator for salary talks. For career planning, free AI coaching through

Coach by CareerVillage provides guidance to explore career paths, while Google's Career Dreamer maps out potential trajectories based on your interests and experience. Perhaps most impressively, AI can compress what used to take forty hours of career planning research into just one-two minutes, analyzing your skills and identifying gaps while recommending specific courses or certifications to bridge them. The key insight for parents is that these tools don't replace human judgment—they enhance it, giving your children the confidence and data they need to make strategic career moves. Try these prompts.

Help me structure a salary negotiation email that emphasizes my value while remaining professional. I'm currently making $X, the market rate appears to be $Y, and my key achievements include [list two-three specific accomplishments with metrics]. The role is [job title] at [company type].

Based on my current role as [job title] with skills in [list three-to-four key skills], what are three potential career advancement paths? For each path, tell me what additional skills I'd need to develop and suggest specific steps to get there within two-three years.

Analyze this job description for my target role: [paste job description]. Compare it to my current background: [brief summary of experience]. What are the top five skills I'm missing, and what specific online courses, certifications, or experiences would best help me close these gaps?

- **Building Professional Portfolios**—AI tools like Canva's Magic Studio make it easy for anyone to create impressive portfolios, even without design or technical skills. These platforms help students turn basic school projects into professional showcases—whether it's using Behance for creative work, GitHub for coding projects, or Storydoc AI for business presentations. The secret is using AI to handle the heavy lifting while adding your own personal stories and authentic voice to make the portfolio truly yours. Here are a couple of powerful prompts to use in LLMs:

> Take this school/internship project and help me describe it in professional terms for my portfolio: [describe project]. Focus on the skills I used, challenges I solved, and any measurable outcomes. Write it as a compelling case study that shows my problem-solving abilities.

> I need to write a portfolio description for [type of project - website, research paper, group project, etc.]. Here are the basic details: [provide context]. Help me craft a two-three paragraph description that highlights my technical skills, creative thinking, and impact, using professional language that would impress potential employers.

Making It Work for Your Family

- **Implementation Guide for Parents**—Getting started with AI job search tools requires patience and practice. Most families need several weeks just to learn how to write effective prompts and review AI output properly before seeing real improvements. Begin with free tools like LinkedIn, Indeed, and ZipRecruiter to avoid financial pressure while you and your child figure out which approaches work best, keeping in mind that the first few resumes and cover letters will likely need multiple revisions. The timeline should be flexible since some kids pick up AI prompting quickly while others need more time to feel comfortable, and getting actual interviews can take months depending on the job market and field. The most realistic expectation is that AI tools will make the process more organized and less overwhelming, but they won't magically speed up hiring timelines since success comes from consistently using these tools over time rather than expecting immediate results.

- **Keeping It Real and Authentic**—The biggest risk with AI tools is creating generic applications that sound robotic, so parents must teach their children to always treat AI output as a starting point that requires significant personalization with their own voice, specific examples, and genuine experiences. Fact-checking is crucial since AI can make up details like job titles, dates, or company information, meaning every resume and cover letter needs careful review to ensure all claims are truthful and accurately represent their actual background. The goal is using AI for efficiency while maintaining authenticity by adding personal anecdotes, adjusting tone to match how they naturally speak, and ensuring the final application genuinely reflects who they are rather than what AI thinks they should be.

Chapter Eight

Teaching AI Safety

Just as you once taught your child to look both ways before crossing the street and make good choices online, you now face the challenge of guiding them through an AI-powered world. The difference is that while you learned these other life skills through your own childhood experiences, AI is largely new territory for most parents. You might feel overwhelmed by headlines about AI dangers or worried that you're not equipped to teach your child about something you're still learning yourself. But here's the good news: you don't need to become an AI expert to raise children who use these powerful tools responsibly. What you need are practical strategies, age-appropriate guidelines, and the confidence to have ongoing conversations with your child about making smart choices with AI. This chapter will give you exactly that. You'll get a clear roadmap for teaching your family to use AI's incredible benefits while staying in control, protecting your privacy, and keeping the critical thinking skills that make us human. By the end of this chapter, you'll have simple tools to help your child become not just AI-literate, but AI-wise.

Current AI Risks Your Family Should Know About

Teaching your child about AI is like teaching them to cross the street safely. You need to understand the basics first. There are four key areas every parent should know about. Privacy and data protection is about keeping your family's personal information safe. AI systems collect and use everything from your child's photos and messages to their homework and online searches. Fairness and bias means understanding that AI isn't fair to everyone. These systems often treat different groups of people unfairly, like assuming boys are better at math or that people of certain races are more likely to be criminals. Understanding How AI works means knowing how these systems make decisions. While AI companies are getting better at explaining their technology, much of what happens inside these systems is still a mystery. Safety and security is about protecting your family from AI dangers. Problems with AI have increased by over half in just one year, reaching 233 reported cases in 2024 according to the AI Incidents Database. (Artificial Intelligence Index Report 2025, p.167).[1]

The dangers your family faces today are real and happening right now. AI systems often make things up or hallucinate, making things sound completely true but are actually false. For example, ChatGPT has created fake research papers and made-up news stories that never happened. These systems also show unfair treatment toward different groups of people. Even the most advanced AI systems have shown patterns of bias in associating certain words with specific races or genders, according to studies. Privacy is becoming a bigger problem as fewer websites allow AI companies to use their information for training. This went from about 5-7% of websites to 20-33% in just one year. (Artificial Intelligence Index Report 2025, p. 163). Security problems now include fake videos and audio recordings that look and sound real. During 2024's elections around the world, AI-generated lies appeared on ten different social media

---

1. Artificial Intelligence Index Report 2025, p167

platforms across fifteen countries. (Artificial Intelligence Index Report 2025, p. 165).[2]

These problems aren't something that might happen in the future. They're affecting families today. A teenage girl in Texas was harassed when a classmate used AI to create fake nude photos of her from her social media pictures. Read about it in the June 25, 2024 article "Girl, 15, calls for criminal penalties after classmate made deepfake nudes of her and posted on social media" published in the U.S. Edition of the *Independent* on independent.co.uk.[3] And a 14-year-old boy sadly took his own life after talking to an AI chatbot that gave him harmful advice instead of helping him. Read about it in the October 23, 2024 article titled "Boy, 14, fell in love with 'Game of Thrones' chatbot — then killed himself after AI app told him to 'come home' to 'her': mom" by Emily Crane in the *New York Post* on nypost.com.[4] A woman in England was wrongly accused of shoplifting by an AI face recognition system and was embarrassed in public before anyone realized the mistake. Read about it the May 25, 2024 article by James Clayton titled "I was misidentified as shoplifter by facial recognition tech" on bbc.com.[5] The answer isn't to ban AI from your home but to help your children learn how to use it safely and smartly. When you help your family understand these four key areas and current dangers, you're preparing them to get the benefits of AI while avoiding the risks. The goal is raising kids who use AI as a tool, not a crutch and keep their ability to think independently.

2. Artificial Intelligence Index Report 2025, p165

3. https://www.independent.co.uk/news/world/americas/elliston-berry-dee pfakes-social-media-b2566806.html

4. https://nypost.com/2024/10/23/us-news/florida-boy-14-killed-himself -after-falling-in-love-with-game-of-thrones-a-i-chatbot-lawsuit

5. https://www.bbc.com/news/technology-69055945

Teaching Your Child to Stay in Control of AI

Reid Hoffman, a tech expert who helped create LinkedIn, gives parents a simple way to think about AI that he calls the "Techno-Humanist Compass." Think of it as your family's GPS for navigating AI safely. The first rule is that AI should work FOR you, not ON you. ("Superagency: What Could Possibly Go Right with Our AI Future," by Reid Hoffman).[6] This means AI should help your child do what they want to do, not try to change their mind or manipulate them. For example, AI should help with homework research, not convince them to buy things or believe certain ideas. The second rule is that technology should make us more human, not less. AI should help your child be more creative, learn faster, or solve problems better, but it shouldn't replace their thinking, friendships, or decision-making. Transparency matters means teaching your child to always ask, "How did you figure that out?" when AI gives them an answer. Just like you'd want to know how a stranger knew something about your family, your child should question how AI knows what it claims to know. Finally, be active participants, not just users, meaning your family should think about and discuss AI, not just accept whatever it says or does.

The most important rule for your family is what experts call "Human in the Loop," but you can think of it as "People Always Decide." This means that no matter how smart AI seems, a human person should always make the final choice about important things. Show your children that AI is like a very advanced calculator or research assistant. It can give them information and suggestions, but they are always the boss. Just like you wouldn't let a GPS drive your car for you, your child shouldn't let AI make decisions for them. This is

---

6.   *Superagency: What Could Possibly Go Right with Our AI Future,* by Reid Hoffman

especially important for things like what to believe, how to treat other people, what to write for school, or any choice that affects their life or someone else's life. Your child should learn to think of AI as a powerful tool, like a hammer or a computer, not as a teacher or authority figure who knows what's best for them.

Building this healthy relationship with AI starts with understanding what AI actually is. Scientists sometimes call AI a "fancy pattern-matching system" or use the term "stochastic parrot," which sounds complicated but just means AI is really good at copying and mixing together things it has seen before, like a parrot that repeats words without understanding what they mean. Your child should learn that AI doesn't truly understand anything the way humans do. It can't feel emotions, doesn't have personal experiences, and doesn't know right from wrong unless humans teach it. This isn't meant to scare your child away from AI, but to help them approach it with healthy curiosity and skepticism. Teach them to think, "That's interesting, but let me check that," instead of, "AI said it, so it must be true." When your family follows these guidelines, AI becomes a powerful helper that makes your child smarter and more capable, while they stay firmly in control of their own thoughts, decisions, and life.

Age-Appropriate Guidelines

- **Elementary Age (six to ten): Building the Foundation Together**—With young children, think of AI like introducing them to a new pet that needs careful watching. Stay with your child when they use AI tools, just like you would when they're learning to ride a bike. Focus on fun, creative activities that spark their imagination, like using AI to help create stories, draw pictures, or answer simple questions about animals or space. The most important lesson at this age is teaching them that "AI can make mistakes, just like people do." When an AI tool gives a wrong answer or creates something silly, use it as a learning moment to say, "See? Even computers get things wrong sometimes." Establish the golden rule early: "Always think for yourself first." Before asking AI anything, have your child tell you what they think the answer might be. This builds their confidence in their own thinking while showing them that AI is just one source of information, not the ultimate authority.

- **Middle School (eleven to thirteen): Developing Smart Habits**—As your child becomes more independent online, it's time to teach them to be AI detectives. Start with fact-checking basics: When AI gives them information for a school project, show them how to verify it by looking at two or three other reliable sources like their textbook, library books, or trusted websites. Help them understand that AI isn't neutral by showing examples of how it might give different answers about the same topic to different people, or how it might reflect unfair opinions about certain groups. Privacy becomes crucial at this age, so establish clear rules about what's safe to share with AI systems (general homework questions are okay, but personal family information, passwords, or private thoughts are not). Most importantly, help them understand the difference between using AI as a research assistant versus letting it do their work for them. AI can help them brainstorm ideas or understand difficult concepts, but their final project should be their own thoughts and words.

- **High School (fourteen to eighteen): Preparing for the Future**—Teenagers are ready for deeper conversations about AI's role in society and their future lives. Start discussing real-world issues like how AI affects jobs, privacy, and fairness in society. These aren't just abstract concepts anymore; they're preparing for a world where AI will be part of their college applications, job interviews, and daily work life. Help them develop strong academic integrity habits by creating family rules about AI use in homework: They should always tell their teachers when they've used AI for help, never submit AI-generated work as their own, and understand that using AI to cheat hurts their own learning. Encourage them to join conversations about AI policies in their school and community, because their generation will be making decisions about AI rules for the rest of their lives. Finally, help them think about their future careers by exploring how AI might change the jobs they're interested in. Will AI make their dream job easier, more creative, or completely different? This isn't about scaring them, but about helping them prepare to thrive in an AI-powered world by developing skills that complement, rather than compete with, artificial intelligence.

Practical Implementation

Help your child develop good AI habits by teaching them three simple questions to ask every time they use AI. Before using any AI tool, they should ask, "What am I trying to achieve?" This helps them stay focused on their real goal instead of getting distracted by whatever AI suggests. For example, if they need help with a math problem, their goal is to understand the concept, not just get the right answer. While using AI, they should ask, "Is this working as expected?" If AI gives confusing answers, goes off-topic, or suggests something that doesn't feel right, they should stop and try a different approach or ask for help. After they finish using AI, they should reflect: "Did this help me or harm me? Did it help or harm others?" This last question is especially important because AI decisions can affect classmates, family members, or even strangers. For instance, if AI helps them write a school presentation about a controversial topic, they should consider whether their presentation might hurt or mislead their classmates.

Just like you have rules about TV time, chores, and bedtime, your family needs clear rules about AI use. Start by setting boundaries around when and how long your child can use AI tools, treating it similar to screen time limits. Establish transparency rules for schoolwork: Your child must tell you and their teachers whenever they use AI for homework help, and they should save their conversations with AI to show their thought process. Create privacy protection rules that are easy to remember, like "Never tell AI your full name, address, phone number, or family secrets." Many families find it helpful to have regular conversations about AI experiences, similar to how you might naturally ask about their day at school. These casual check-ins can help you notice any concerns while also celebrating moments when AI genuinely helps your child learn something new. Some parents find it useful to write down their family's AI guidelines and keep them in a visible spot as a gentle reminder, though every family will find their own approach to staying consistent with their values around AI use.

Help your child to recognize when AI might be causing problems, just like you'd teach them to recognize when a friend is being a bad influence. Be suspicious of AI systems that won't explain how they came up with their answers, especially for important decisions. If AI seems to know personal information about your family that you never shared, that's a red flag that your privacy might be compromised. Be very careful with AI tools that make important choices without involving humans, like automatically sending emails, making purchases, or giving medical advice. Finally, teach your child to notice when AI says unfair things about different groups of people, like suggesting that certain races, genders, or religions are better or worse than others. When your child spots these warning signs, they should stop using that AI tool and talk to you about what happened. Remember, the goal isn't to make your child afraid of

AI, but to help them become smart, careful users who can enjoy AI's benefits while avoiding its risks.

Teaching Your Child to Use AI Wisely While Building Real Skills

When it comes to schoolwork, help your child understand the difference between using AI as a thinking partner versus letting it do their work for them. AI can be great for brainstorming ideas, like helping them think of different angles for a history project or explaining a difficult science concept in simpler terms. However, the final work should always be their own thoughts, written in their own words. Teach them to always tell their teachers when they've used AI for help, just like they would mention if they got help from a tutor or parent. This isn't about getting in trouble; it's about being honest and showing their learning process. Work with your child's teachers to understand their school's AI policies, because different teachers and schools have different rules. Some might encourage AI use for research, while others might not allow it at all. The key is building your child's own skills and knowledge rather than becoming dependent on AI. Think of it like using a calculator in math class: it's a helpful tool, but your child still needs to understand the underlying concepts.

Protecting your family's privacy in an AI world requires understanding how these companies collect and use personal information. AI systems learn from massive amounts of data, including social media posts, search histories, and even conversations people have with AI chatbots. Help your child understand that everything they share online becomes part of their "digital footprint," a permanent record that follows them throughout their life. Help them make smart decisions about what to share with AI systems: General homework questions are usually fine, but personal family details, passwords, or private thoughts should stay private. When your child encounters AI-generated content online, whether it's a news article, video, or social media post, teach them to be a detective. Show them how to check if information is real by looking at multiple trusted sources, checking the date to make sure the information isn't outdated, and being extra careful with content that seems designed to make them angry or scared. Help them develop healthy skepticism without becoming paranoid by teaching them to think, "This might be true, but let me check," rather than automatically believing or dismissing everything they see. With deepfakes and AI-generated videos becoming more common, remind them that even seeing isn't always believing anymore, and it's always okay to ask you or a teacher if something seems suspicious.

Just as you successfully taught your child to navigate busy streets, make good choices online, and develop other essential life skills, you now have the tools to guide them through our AI-powered world with confidence. Remember, this isn't a one-time conversation, but an ongoing journey of discovery together. Some days your child will amaze you with their thoughtful questions about AI, while other days you'll need to remind them of the basics, and that's perfectly normal. The strategies in this chapter aren't just about keeping your family safe; they're about raising a generation that will shape how AI develops and how society uses these powerful tools. When you teach your child to think critically, stay in control, and use AI as a helpful tool rather than a replacement for human judgment, you're not just protecting them. You're helping create future leaders, innovators, and decision-makers who will ensure AI serves humanity's best interests. Trust yourself, start with small steps, and remember that your common sense and family values are exactly what your child needs to thrive in an AI-enhanced world. The future belongs to those who can work alongside AI while staying true to what makes us human, and you're giving your child that invaluable gift.

Chapter Nine

Your Child's AI Future

Your child is growing up during one of the most significant technological shifts in human history, but unlike previous changes that unfolded over decades, this AI transformation is happening at unprecedented speed. Just as the internet transformed how we communicate, learn, and work over the past thirty years, AI will reshape virtually every aspect of daily life in the decades ahead, but in years, not decades. The AI capabilities doubling every few months means that the tools your child learns this year may be completely transformed by next year, requiring families to be more adaptive and continuously learning alongside the technology.

Think of AI becoming as common and essential as smartphones are today, but far more powerful and deeply integrated into daily life. Your child will likely have personal AI assistants that know their learning style, help with homework, manage their schedules, and even provide emotional support during difficult times. These won't be the clunky chatbots we see today, but sophisticated companions that understand context, remember conversations, and adapt to individual needs. More importantly, instead of juggling separate apps for creative projects, homework help, college prep, and productivity as described throughout this book, your child will likely have unified AI companions that seamlessly handle all these needs while learning their preferences across every domain of their life.

The job market your child enters will look dramatically different from today's workplace, and AI fluency won't be optional. It will be expected. Just as employers today assume workers can use email and basic computer programs, future employers will expect candidates to seamlessly navigate AI tools and understand how to leverage them effectively. This shift explains why AI literacy is rapidly becoming formalized in education systems worldwide. Just as computer literacy classes became standard curriculum in the 1990s, expect AI instruction to become as fundamental as reading, writing, and arithmetic in your child's school experience. While some traditional roles may disappear, entirely new categories of work will emerge that we can barely imagine now. It's critical that your child learns to collaborate with AI systems now, using them to enhance their natural human abilities rather than replace their thinking. Those who develop this skill early will have tremendous advantages, while those who avoid AI entirely may find themselves at a significant disadvantage when competing for jobs and opportunities.

The democratization of such powerful capabilities creates both incredible opportunities and unexpected challenges. From professional-quality video creation to sophisticated writing assistance to advanced data analysis, your child will have access to tools that were once available only to experts with expensive equipment and years of training. This levels the playing field for creativity and innovation, but it also means that traditional gatekeepers and skill hierarchies may be disrupted in ways we're just beginning to understand. The same technology that empowers your child to create amazing projects might also reshape entire industries and professions in surprising ways.

As AI becomes more prevalent, families will face new decisions about privacy, screen time, and digital relationships that go far beyond current concerns about social media or gaming. The AI systems of the future will know intimate details about our children's lives, learning patterns, and preferences, evolving from simple tools to something approaching collaborative partners or even emotional companions. This creates both incredible opportunities for personalized education and support, as well as important questions about data protection, healthy boundaries, and the nature of human-AI relationships as they become more sophisticated and persistent in our children's lives.

The transformation extends far beyond individual family use. Your child will inherit a world where AI reshapes healthcare systems, transportation networks, government services, and social institutions. The global competition for AI leadership means that the choices made by governments and companies today will shape the digital world your child inhabits tomorrow, influencing everything from the privacy protections they enjoy to the opportunities available to them.

The most important thing you can do is to stay engaged and curious alongside your child. You don't need to become an AI expert, but maintaining open conversations about technology, modeling responsible digital habits, and helping your child think critically about the tools they use will serve them well. Remember that this is a continuous journey. What you learn together this year will need updating next year as capabilities advance and new challenges emerge. The future won't be about humans versus machines, but about humans and machines working together to solve problems and create opportunities that neither could achieve alone. By helping your child develop both technological fluency and strong human values, you're preparing them not just to adapt to this future, but to help shape it in positive ways. Your child's generation won't just be users of AI. They'll be the architects of how AI integrates with human society, making the foundation you build together today more important than ever.

Final Thoughts

Artificial intelligence is no longer something to ignore or leave to the tech world. It's woven into our daily lives and will shape our children's future careers. Understanding AI is essential, because future employers won't just hope our children know how to use AI—they'll expect and depend on it. As parents, we have a powerful role to play, not as experts, but as guides willing to learn alongside our children.

Throughout this book, I've aimed to show that AI isn't something to fear. When used responsibly, it can unlock creativity, enhance learning, and even save time and money for your family. The key isn't knowing everything about AI. Instead, it's about asking thoughtful questions, setting appropriate boundaries, and staying engaged as your child explores these tools.

My own turning point came when I watched my son use AI to discuss a book he was reading. He engaged in meaningful conversation about the text while maintaining his own critical thinking. This showed me that children can harness AI's power without losing their independence of thought.

Remember, you don't have to get it perfect, and you don't need to navigate this alone. Simply start the conversation, stay curious, and keep showing up for your child's digital journey. The future belongs to families who embrace exploration together.

I hope this book has helped you feel more confident about integrating AI into your family's life, regardless of your technical background. Whether for learning, creativity, budgeting, college preparation, or career development, AI can become a valuable tool for your family when approached with both care and curiosity.

Get Involved—Free AI Learning Resources

Beginner-Friendly Courses and Tutorials

- **OpenAI Academy** (academy.openai.com)—Interactive webinars and self-paced tutorials covering ChatGPT fundamentals and advanced prompt engineering techniques

- **Anthropic AI Fluency** (anthropic.com/ai-fluency)—Focused training on collaborating with AI systems effectively, ethically, and safely

- **Google AI Essentials** (grow.google/certificates/en_ca/ai-essentials /)—Comprehensive fifteen-hour certification program covering AI basics and practical applications

Technical Skill Building

- **Microsoft AI Learning Hub** (learn.microsoft.com/en-us/ai/)—H ands-on tutorials and documentation for integrating AI into applications and workflows

Industry Insights and Data

- **Stanford AI Index Report** (hai.stanford.edu/ai-index)—Annual comprehensive analysis of global AI trends, development patterns, and societal impact

- **Section AI Webinars** (sectionai.com)—Regular expert-led sessions on current AI applications and business strategies

Networking and Events

- **Luma AI Events** (lu.ma/ai)—Curated AI meetups and conferences, primarily Bay Area-based with select virtual options

- **Cerebral Valley Events** (cerebralvalley.ai/events)—Silicon Valley's premier AI community events and networking opportunities

AI Timeline

Here is a timeline of Artificial Intelligence Evolution (1940s to the Present) of major events on how we got here:

- 1943: Warren McCulloch and Walter Pitts propose the first mathematical model of a neural network.

- 1956: John McCarthy organizes the Dartmouth Summer Research Project on AI, where the term "Artificial Intelligence" is coined and the field formally begins

- 1958: Frank Rosenblatt builds the Mark 1 Perceptron machine

- 1966: Joseph Weizenbaum creates ELIZA, an early natural language processing chatbot that could simulate conversation.

- 1980: Expert systems become commercially viable and widely used in industry. Digital Equipment Corporation deploys XCON (R1), a rule-based configurator.

- 1986: Backpropagation algorithm is rediscovered, allowing for more effective neural network training.

- 1995: Dr. Richard Wallace creates ALICE, an advanced chatbot.

- 1997: IBM's Deep Blue defeats world chess champion Garry Kasparov.

- 2007: iPhone debuts, bringing computing power and AI applications to consumers' pockets.

- 2010: Mustafa Suleyman, Demis Hassabis, and Shane Legg found DeepMind.

- 2011: IBM's Watson defeats champions on Jeopardy!

- 2012: AlexNet, a deep convolutional neural network wins the ImageNet vision challenge by a huge margin.

- 2014: Google acquires DeepMind for $650 million.

- 2015: Elon Musk, Sam Altman, and others co-founded OpenAI.

- 2016: DeepMind's AlphaGo program defeats Go world champion Lee Sedol.

- 2018: OpenAI releases its first Generative Pre-trained Transformer (GPT-1).

- 2018: BERT model by Google demonstrates breakthrough performance in natural language processing.

- 2020: OpenAI releases GPT-3, a 175-billion-parameter language model.

- 2022: OpenAI releases DALL-E 2, bringing text-to-image generation to the public.

- 2022: OpenAI releases ChatGPT, reaching 1 million users in less than a week.

- 2023: OpenAI releases GPT-4.

- 2023: Meta open-sources LLaMA-2.

- 2023: President Biden signs an Executive Order on AI safety and regulation.

- 2023: Anthropic's Claude 2 achieves new levels of factual accuracy and context length.

- 2024: OpenAI demonstrates GPT-4o, showing advanced multimodal capabilities.

- 2025: AI systems become increasingly integrated into daily life and professional domains.

AI Dictionary

- **Agentic AI**: Systems capable of setting goals and taking actions on their own to achieve them, functioning more autonomously with minimal human oversight.

- **Artificial General Intelligence (AGI)**: The hypothetical ability of an AI system to understand, learn, and apply knowledge across a wide range of tasks at a human level or beyond.

- **Artificial Intelligence (AI)**: The broad term coined by John McCarthy in 1956 describing the field of making machines intelligent.

- **Benchmarks**: Standardized tests used to measure and compare AI system performance across various dimensions like accuracy, reliability, and safety.

- **Context Window**: The amount of information (measured in tokens) that a model can "remember" and process at one time, acting like its short-term or working memory during a conversation or task.

- **Deep Learning**: A term describing advanced neural networks with multiple layers (originally "deep neural networks"), popularized as a rebranding of neural network approaches when they returned to favor.

- **Expert Systems**: A type of rule-based AI popular in the 1980s that attempted to encode the specialized knowledge of human experts into domain-specific systems.

- **Fine-Tuning**: The process of training a pre-trained model on specific data to make it better at particular tasks.

- **Foundation Models**: Base AI systems that can be adapted for various applications, which companies like Inflection create.

- **Generative AI**: Models that can create new content (text, images, etc.) rather than just analyzing existing data.

- **Hallucinations**: When models generate false or misleading information not grounded in their training data.

- **Iterative Deployment**: OpenAI's approach of releasing AI capabilities incrementally to allow society time to adapt, while gathering feedback from real-world usage.

- **Large Language Models (LLMs)**: The foundation of modern conversational AI systems like ChatGPT. These are machine learning constructs designed for language processing tasks that use neural network architecture to process and predict patterns in language.

- **Machine Learning**: Systems that improve their performance through experience rather than explicit programming, a broader category that includes neural nets.

- **Model Context Protocol (MCP)**: Open, standardized protocol designed to connect AI models—especially large language models (LLMs)—to external data sources, tools, and services in a secure and consistent way.

- **Multimodal Learning**: AI systems that can process multiple types of inputs like text, audio, images, and video together rather than just text.

- **Neural Networks**: The underlying architecture of LLMs consisting of multiple layers of nodes that perform interconnected computations.

- **Neurosymbolic AI**: Systems that integrate neural networks with symbolic reasoning based on explicit, human-defined rules and logic.

- **Parameters**: Described as "tuning knobs" in neural networks—hundreds of billions of these determine the strength of connections between nodes in modern large models.

- **Prompt:** A question, instruction, or piece of information you give to an AI system—like typing a request into ChatGPT—that tells it what kind of response you want, and the more specific and clear your prompt is, the better and more accurate the AI's answer will be.

- **Reasoning Model**: A type of AI, often a large language model, specifically designed to solve complex problems by breaking them down into smaller steps and working through them in a logical, human-like way.

- **Reinforcement Learning**: A technique where AI models learn by receiving feedback (rewards or penalties) based on their actions, used prominently by DeepMind.

- **Reinforcement Learning with Human Feedback (RLHF)**: A technique where humans rate AI outputs to help models learn which responses are most desirable.

- **Retrieval Augmented Generation (RAG)** – AI technique that improves language model responses by combining real-time information retrieval from external sources with text generation.

- **Rule-based Computing/Symbolic AI**: An early approach to AI that relied on explicit programming and predefined rules, championed by Marvin Minsky and others who opposed neural networks.

- **Stochastic Parrot**: A term coined by Emily M. Bender to describe LLMs that generate impressive-sounding language but without true understanding.

- **Tokens**: The basic units that LLMs process—words or fragments of words. The model analyzes associations between tokens during training.

- **Transformer**: A neural network architecture introduced by Google researchers that revolutionized language models by efficiently processing relationships between words.

- **Vector Database** – Specialized database designed to store, index, and search high-dimensional numerical representations called vector embeddings of data such as text, images, or audio, typically generated by machine learning models.

Acknowledgements

Thank you to Peter Zwart for encouraging me to write this book. I'm deeply grateful to my husband of over twenty-seven years, whose thoughtful contributions and experience in applying AI-based capabilities in marketing helped shape my approach to practical, real-life applications for parents. From brainstorming ideas to offering guidance and encouragement, your input was invaluable. You kept our ravenous teen boys well-fed and grounded while I attended countless AI-related events across the Bay Area to research and gather material. This book is better because of you.

About the author

James McConihe has worked in technology for over thirty years with companies including AT&T, Apple, MacUser Magazine (Ziff Davis Publishing), and Relax Technology. He also served for more than five years as both Vice President and President of the Educational Fund at his son's elementary and middle school. James lives in the San Francisco Bay Area with his husband and their two sons.

ABOOKS

ALIVE Book Publishing and ALIVE Publishing Group
are imprints of Advanced Publishing LLC,
3200 A Danville Blvd., Suite 204, Alamo, California 94507

Telephone: 925.837.7303
alivebookpublishing.com

www.ingramcontent.com/pod-product-compliance
Lightning Source LLC
Chambersburg PA
CBHW022034090426
42741CB00007B/1057